LEVERS

Chris Ollerenshaw and Pat Triggs
Photographs by Peter J. Millard

Contents

Words that appear in the glossary are printed in
boldface type the first time they occur in the text.

Gareth Stevens Publishing
MILWAUKEE

What's in my toy box?

Some of the toys in this toy box have been put into separate containers. But the lid of one container is stuck. How would you get it open?

A spoon would be useful.

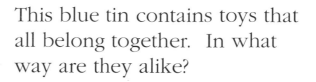

This blue tin contains toys that all belong together. In what way are they alike?

Look at the red and yellow toy traffic signal. Then look at the spoon prying off the container's lid. Is the up-and-down movement of the signal arm like the movement of the spoon opening the lid?

The arm of the signal and the spoon lifting the lid are both **levers**. All the toys in the blue container have something to do with levers.

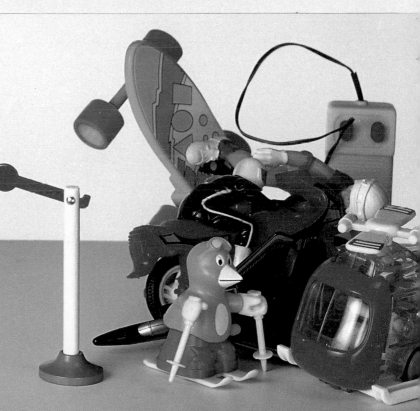

How does a spoon become a lever?

A spoon is a tool designed for eating things like soup, cereal, and ice cream. But the shape of a spoon also makes it a useful lever for other purposes.

When you put one end of a spoon under a lid and push down on the other end, you are using the spoon as a lever. When you push down on one end of the spoon, the other end goes up, just like a seesaw. The point where the spoon **pivots** is called a **fulcrum**. All levers have a fulcrum.

Collect some levers like these and discover what working with levers feels like. Notice where the fulcrum is each time you try a different lever. Make a lever to help you lift a pile of books. What kind of materials do you need?

Collect a group of objects that you think are levers.

There are types of levers in your body. Where do you think they are?

Body levers

Humans and other animals have many useful fulcrums or joints in their bodies. Every time you eat or walk across a room, you are using the levers in your body. Your arms, legs, and jaw – all these parts of your skeleton – contain levers. The fulcrum is the hinged joint that allows the bones to swivel but keeps them together.

Each of our joints holds two levers together. Look at this jaw bone of a lion. Can you see how the bones are hinged together?

Thousands of years ago, people began to see how levers could be useful to them. This hinged lid on a purse is about 1,300 years old.

We use hinges in many ways. How many objects with hinges can you find around your home?

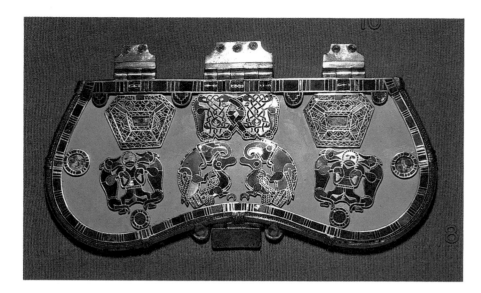

Why are levers useful?

A lever is a machine that helps us accomplish tasks. When we think of machines, we usually think of things such as computers or race cars. But when you used the spoon as a lever, it became a simple machine accomplishing a task.

Some other simple lever machines are shovels and oars. Levers make work easier. They don't make you stronger, but they use your strength effectively so the job becomes easier. But that's not all levers can do.

You can discover what other uses levers have by making some yourself. You will need: a piece of thick cardboard, some strips of thin cardboard in various colors, thumbtacks, and paper fasteners. These materials will be used in the experiments on the following pages.

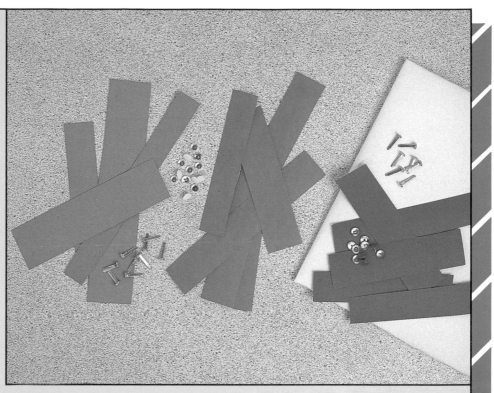

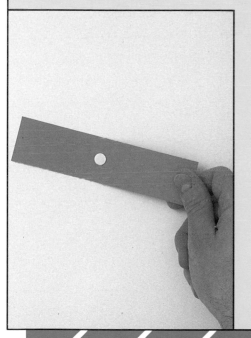

Tack one strip of thin cardboard to the thick cardboard. Move one end of the strip with your hand. Does the tack allow you to move the strip up and down? If it does, the tack is a fulcrum, and you have made a lever!

Notice how the ends of the strip are moving. If you make one end of a lever go down, the other end goes up. Try moving the tack to different positions on the strip. What difference does this make?

Making levers

Now add a second strip like the blue one in the picture. Punch a hole in both pieces of cardboard where you want the pieces to join. Use a paper fastener to hold them together. Try moving the end of the strip you have added. Watch how it moves. Watch how the first strip moves. You have made two levers.

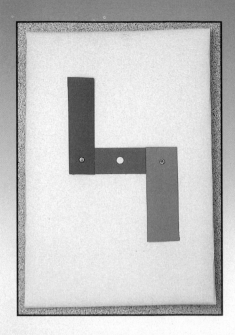

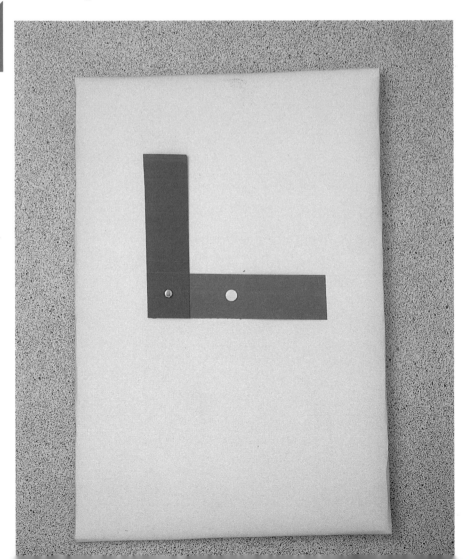

Now add a third strip like this green one. Try the movement again. How many levers have you made now?

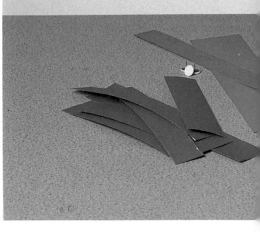

To find out how many levers you have made, look carefully at the green strip you have just added. Is it swiveling on the paper fastener? You can check whether it is by looking at the ends of all the strips. If your third strip is not pivoting on the paper fastener, it will only be able to move in the same direction as the end of the strip that is tacked to the board.

If the green strip isn't able to swivel, the paper fastener is not a fulcrum, and you have not made a third lever.

On the green strip in the picture, the fastener is not a fulcrum. There are still only two levers.

Changing direction

Before you can turn the green strip into a lever, you have to find a way to control how that strip moves. Put some mapping pins along each side of your strip. (Make sure you don't pin the strip down so that it can't move.)

Now move your blue lever up and down. Watch your green strip. It has to move in the pathway made by the pins. To do that, it has to pivot on the paper fastener. When it does that, you'll have three levers all moving in different directions. When you pull the blue lever down what does the green one do? What happens when you push the blue lever up? Use mapping pins to make different pathways for the green strip. What happens? Do all the pathways work equally well?

Here's a challenge. How many strips can you link together to make a sequence of levers? Try using longer or shorter strips.

Here's a pop-up toy. From the front, you cannot see how it works. But you now know enough about levers to imagine how it works. Try making your own pop-up toy.

Turning small movements into bigger ones

This toy signal uses a lever. Use two strips of thin cardboard and your board to make a lever that works like the signal. With your fingers near the pivot fastener, move the signal arm up and down. Notice how you only have to move your finger a little to make the other end move a bigger distance. What happens if you put short, middle-sized, and long strips on the same pivot? When you move your fingers along the lever to push or pull in different places, how does the speed of your fingers compare with the speed of the movable end of the lever?

The designer of this machine for enlarging drawings knew that levers can turn small movements into bigger ones. As one pencil traces the small drawing, the other pencil makes an exact copy, but in a larger size.

Making things move

With this toy signal, all the levers are linked together in a sequence, or series.

Make a model, as shown, to discover how levers work together to move the signal arm.

Some wastebaskets have a foot pedal that lifts a lid on top. They work almost exactly the same way as the signal. Use your tacks, fasteners, and strips to show how they work. You will not have to change your model much. Show how when the pedal goes down, the lid goes up.

Use your model to show how this chicken pecking toy works. The signal, the wastebasket, and this toy all work in the same way.

Making work easier

Levers help you put more force on objects that are difficult to move or lift. They can make something move in a different direction. And they can turn a small movement into a larger one.

Look at this toy. When you pull down on the string, its arms and legs move in different directions. You need to pull down only a little ways to make the toy's hands and feet move a long way.

Can you guess what is happening at the back of this toy? Where do you think the pivot points are on each lever? What job is the string doing? How is it attached? Try making a model of the movement of this toy using your board and strips. Could you use a cardboard lever instead of string?

Try designing another toy that works like this one.

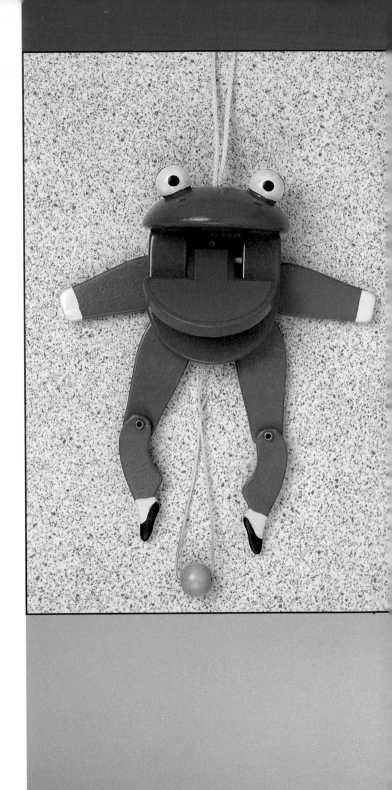

Keeping the balance

Six thousand years ago, people realized that levers could be used to balance objects. With this knowledge, they invented a machine for weighing things. They hung a beam from a cord, making sure there was a central pivot. The Romans used this idea to make a weighing machine called a steelyard.

Can you see the steelyard on the right of this ancient artwork? The meat to be weighed was put into a pan. The weight on the left of the machine was moved along the arm until it balanced with the pan. The weight of the meat was determined by how far the weight had to be moved along the arm. Modern-day doctors measure your weight with the use of scales having a steelyard lever.

Mobiles and moving sculptures have a point of balance on which their levers work.

You can make a mobile with stiff wire (a coat hanger), thread, and objects to hang from it, such as buttons or small plastic toys. How will you make it work? Think about the distance of each object from the point of support to help you make everything come into balance.

Making the difference

The distance from the fulcrum is also important when you are using levers to move things. Sometimes you can pry open the lid of a container with a coin. Sometimes you have to use something stronger and longer, such as a screwdriver.

Why is a screwdriver better than a coin for opening a lid that is stuck very tightly? Why is the screwdriver better at making the force of your hand strong enough to lift the lid?

Naming the parts

When you are prying off a lid, the force of your hand is on one part of the lever. The force applied to a lever is called the **effort**. The force that is going to be moved by the lever is called the **resistance**. If the load to be moved is heavy, the resistance force will be great.

In all lever movements, there needs to be: a resistance to be moved; a fulcrum around which the lever rotates, or pivots; and an effort applied.

Can you point to the resistance, the fulcrum, and the effort in this picture?

Hint: In this case, the fulcrum is between the effort and the resistance.

Look at this dolly. This time, the resistance is between the effort and the fulcrum.

Can you point to the resistance, the effort, and the fulcrum in the lever movement shown above?

Can you think why the resistance is closer to the fulcrum than the effort? Remember what you know about the importance of distances of forces from the fulcrum.

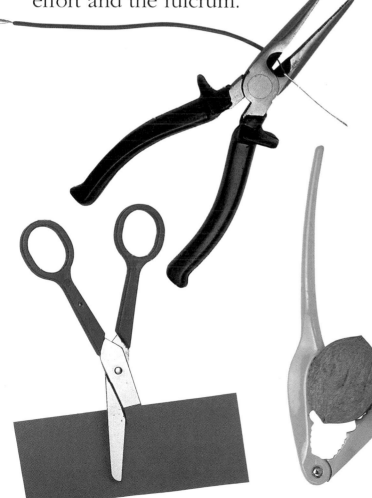

If you put two levers together, you can make useful tools. How do these levers work? Can you see the effort, the resistance, and the fulcrum in each?

All sorts of uses

Try this project to discover another use for levers. Cut out a cardboard circle. Attach a strip lever to the circle with a fastener. Pin the circle to the board. Move your lever up and down in a straight line. Can you see how your straight line movement turns the disc in a circular movement?

Over two hundred years ago, the Scottish engineer James Watt invented a machine that turned back-and-forth movement into circular movement. This was a key part of the steam engine that provided the power for many other machines in factories.

Using levers also allows us to fold objects into a smaller space. Look for the pivots, or hinges, on some objects. Can you see how they work?

Making a moving picture

Each day, whether we realize it or not, we use machines that have levers. Also remember that we use levers in our own bodies when we are doing something as simple as eating or walking.

People who design and make pop-up books use levers to design the movements in their books.

You can make a book using levers by following the plans and instructions on pages 30-31.

A cat hides in a bush. A bird sits in the grass between the bushes. The cat pounces! Too late! The bird flies to safety into the other bush.

You have learned enough to be able to design your own levers to make the cat and bird appear and disappear. The plans will show you how to make the model itself.

28

For a free color catalog describing Gareth Stevens' list of high-quality books, call 1-800-542-2595 (USA) or 1-800-461-9120 (Canada). Gareth Stevens' Fax: 414-225-0377.

The publisher would like to thank Richard E. Haney, Professor Emeritus of Curriculum and Instruction (Science Education) at the University of Wisconsin-Milwaukee for his assistance with the accuracy of the text.

The publisher would like to thank the cover model, David, for his participation.

Library of Congress Cataloging-in-Publication Data

Ollerenshaw, Chris.
    Levers / Chris Ollerenshaw and Pat Triggs.
        p. cm. -- (Toy box science)
    Includes index.
    ISBN 0-8368-1121-6
    1. Levers--Juvenile literature. [1. Levers.] I. Triggs, Pat.
II. Title. III. Series: Ollerenshaw, Chris. Toy box science.
TJ147.045 1994
621.8'11--dc20   94-4886

North American edition first published in 1994 by
Gareth Stevens Publishing
1555 North RiverCenter Drive, Suite 201
Milwaukee, Wisconsin 53212 USA

First published in 1991 by A & C Black (Publishers) Ltd., London. Original text © 1991 by Chris Ollerenshaw and Pat Triggs. Additional end matter © 1994 by Gareth Stevens, Inc. All photographs © Peter J. Millard except p. 6 John Heinrich; p. 7 C. M. Dixon; p. 8 Emrhys Barrell; p. 20 C. M. Dixon; p. 21 Pat Triggs. Model and blueprint by David Ollerenshaw. Illustrations by Dennis Tinkler. Designed by Michael Leaman. Cover photograph © 1994 by Jon Allyn, Creative Photographer.

Series editor: Barbara J. Behm
Cover design: Karen Knutson

Printed in the United States of America

1 2 3 4 5 6 7 8 9 99 98 97 96 95 94

At this time, Gareth Stevens, Inc., does not use 100 percent recycled paper, although the paper used in our books does contain about 30 percent recycled fiber. This decision was made after a careful study of current recycling procedures revealed their dubious environmental benefits.

Pouncing Cat

Photocopy pages 30-31, and then cut out the plans. Trace the plans onto cardboard, starting with the left side of the plans. Notice how the pieces fit together, but don't glue them yet.

Design your levers. The cat and bird should be moved using a lever sticking out of an end wall. Make the cat pounce **DOWN** and the bird fly **UP**.

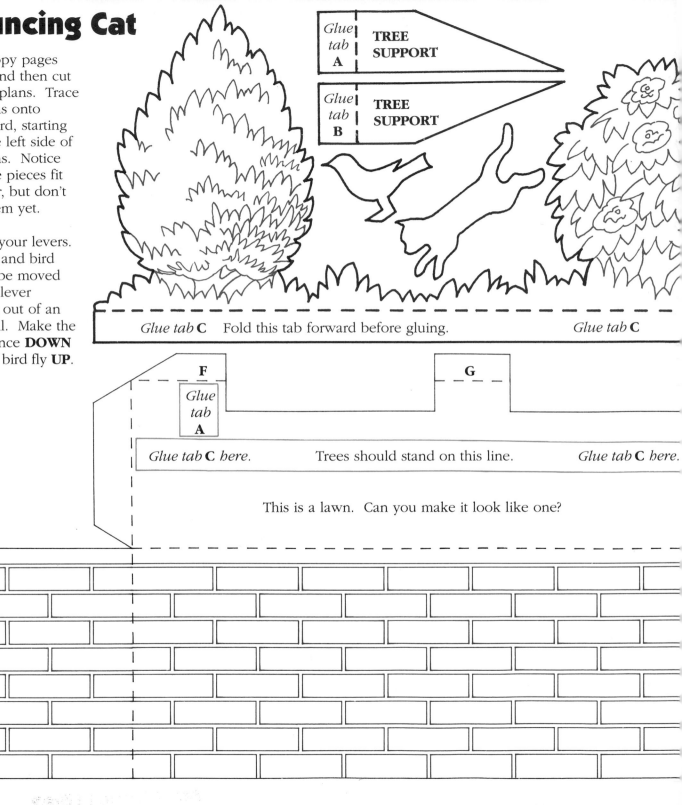

Glue tab A | TREE SUPPORT

Glue tab B | TREE SUPPORT

Glue tab **C**   Fold this tab forward before gluing.   *Glue tab* **C**

F

G

Glue tab A

Glue tab **C** *here.*   Trees should stand on this line.   *Glue tab* **C** *here.*

This is a lawn. Can you make it look like one?

E

D

TREE SUPPORTS keep the trees upright. Trace onto cardboard, cut them out, and glue them into place, matching up the letters. Shorten them if you need to.

Use heavy cardboard for the shed cutout.

CUT along all solid black lines.
FOLD along all dotted black lines.

GREEN lines show where to glue.
RED lines give ideas for decoration.

Fit your levers into this space. If you find this difficult, trace just the outline of the shed cutout (including this oblong) onto heavy cardboard. Then take the cutout of the bushes and use it as a template to draw its outline onto the corrugated cardboard in the correct position. You will then know exactly what movements the animals have to make.

When the levers are finished, cut the heavy cardboard roughly to shape. Trace the shed drawing onto paper and glue it onto the cardboard.

H
Glue tab
B

31

Glossary

effort: the force applied
fulcrum: the point on which a lever is supported
lever: a simple machine used for lifting
pivot: a point on which something turns
resistance: the force that is moved by a lever

Books and Videos

How Everyday Things Work. Chris Cooper and Tony Osman
  (Facts on File)
The Lever and the Pulley. Hal Hellman (M. Evans)
Machines and How They Work. Harvey Weiss (Crowell)
The Simple Facts of Simple Machines. Elizabeth James and Carol Barkin
  (Lathrop, Lee, and Shepard)

Science Discovery for Children (video)
Science Riddles (video)

Index